Teaching
With Purpose

Closing the Research – Practice Gap

D1303678

By John E. Penick and Robin Lee Harris

NATIONAL SCIENCE TEACHERS ASSOCIATION

Arlington, Virginia

NATIONAL SCIENCE TEACHERS ASSOCIATION

Claire Reinburg, Director
Judy Cusick, Senior Editor
Andrew Cocke, Associate Editor
Betty Smith, Associate Editor
Robin Allan, Book Acquisitions Coordinator

Linda Olliver, Cover and Inside Design

PRINTING AND PRODUCTION Catherine Lorrain-Hale, Director
    Nguyet Tran, Assistant Production Manager
    Jack Parker, Electronic Prepress Technician

NATIONAL SCIENCE TEACHERS ASSOCIATION
Gerald F. Wheeler, Executive Director
David Beacom, Publisher

LIBRARY OF CONGRESS CATALOGING-IN-PUBLICATION DATA

Penick, John E.
  Teaching with purpose : closing the research-practice gap / by John E. Penick and Robin Lee Harris.
      p. cm.
  Includes bibliographical references and index.
  ISBN 0-87355-241-5
  1. Science--Study and teaching--Research. 2. Science teachers--Training of. I. Harris, Robin Lee. II.
Title.
  Q181.P3475 2005
  507'.1--dc22
                            2005018612

Contents

Acknowledgment and Dedication

This book grew out of the thinking, practice, and experience of Professor Dorothy Schlitt at Florida State University between 1965 and 1975. Dorothy originated and developed the concept of a research-based teaching rationale, integrating it into her highly effective teacher education program. In subsequent years, John Penick (co-author of this book) developed the rationale concept so that it pervaded every aspect of the science teacher education program he helped develop. Many of John's 29 PhD students (including Robin Harris, his co-author) adopted the concept of a rationale, implementing it in a number of teacher education programs. Over the ensuing 30 years, with institutional constraints, changing demographics, and evolution, the original idea has grown, changed, and been studied. The central concept remains: A personal, research-based rationale can play an essential role in promoting exemplary and effective teaching and professional development.

Dorothy inspired those of us who studied under her. Her words, actions, teaching, and professional life seemed exciting, accurate, deep, purposeful, and, most importantly, successful. Her students both feared and loved her. We knew that here was a marvelous teacher—a truly professional teacher—with insights far beyond any we could conceive, much less describe. Yet, it was not intuitively obvious to us that what we noted and admired in Dorothy was actually planned, coordinated, and based on systematic research, study, and thinking. Little did we know that she changed her ideas regularly and was never content with her current level of success with students, as phenomenal as it seemed to us at the time. We did not understand that her seemingly fully developed ways of thinking and teaching were part of a work in progress. With time though, this became a significant point in our own understanding.

A key ingredient of her personal rationale included constant reflection on her teaching, her teacher education program, her students, and her own ideas and values. Focused and never idle, her reflections led her to additional reading and study, to consider alternate plans and scenarios, to rearrange elements of her teaching, and to try new things. Paradoxically, her seemingly solid and apparently immutable rationale for teaching guaranteed an ever-changing teaching strategy and environment.

The rationales for teaching and for teacher education that each of us developed stem directly from Dorothy Schlitt and have guided us in innumerable ways. We can think of no other aspect of our own professional teacher education that comes even close in terms of impact on us and the students with whom we have worked. Professor Dorothy Schlitt provided us with a strong foundation and a robust framework on which to build. We still see and feel Dorothy's impact in our work daily.

As our careers and the careers of our students have benefited enormously from Dorothy's wisdom and practice, we dedicate *Teaching With Purpose* to Professor Dorothy Schlitt.

John E. Penick
North Carolina State University

Robin Lee Harris
Buffalo State College

About the Authors

John Penick has spent his professional life focusing on how to strengthen the role of the teacher in the classroom. After starting as an inner-city high school teacher, he went on to a 35-year teacher education career at four universities, including North Carolina State University, where he now heads the Department of Mathematics, Science, and Technology Education. A former president of the National Science Teachers Association, Penick earned a bachelor's degree in zoology and chemistry and a master's in biology and community college teaching at the University of Miami. His PhD in science education is from Florida State University.

Robin Lee Harris is an associate professor of science education at Buffalo State College, where her primary research interests concern the development of beliefs and practices in teacher education, assessment practices, professional development, and writing as a tool to thinking. A former middle and high school teacher, Harris has a bachelor's degree in zoology from San Diego State University, a master's in teaching and curriculum from Pennsylvania Sate University, and a doctorate in science education from the University of Iowa, where John Penick was her mentor.

Foreword

Every science teacher strives to maximize student learning. Effective science teachers promote learning by understanding and using research and practices known to work. The education literature contains considerable information about effective teachers and teaching, and researchers have documented much of what has become known as "best practice" in teaching. At the same time, the research literature on human cognition and student learning is growing both stronger and more valid. Yet, many excellent teachers have not learned how to find, evaluate, combine, and use these findings systematically to inform, guide, and ultimately improve their teaching performances. *Teaching With Purpose* is designed to close this gap between research and practice.

Written for teachers, supervisors, teacher educators, and others concerned with providing the best and most effective classroom instruction, *Teaching With Purpose* offers insight into how to develop a research-based plan to improve your work in the classroom. But, rather than attempting to compile all valid research, we describe how to combine your goals and situation with research findings to make you ever more knowledgeable and effective.

Teaching With Purpose requires that you bring your own personal needs, knowledge, experience, energy, taste, tolerance, and learning to bear. If you follow the recommendations and suggestions of this book, you will have a more thorough understanding of your role as a teacher and the impact of that role on student learning. Three key steps are necessary for you to succeed. First, write down your goals, thoughts, understandings, and teaching strategies to begin the process of developing your own "teaching rationale." Second, identify and study the most appropriate and compelling research that might help you improve your success in the classroom. Third, and most importantly, use your personal, research-based rationale in the classroom every day and see your effectiveness increase.

Throughout this book, more than a dozen teachers from across the country, ranging from second-year teachers to 20-year veterans, comment on how developing a research-based rationale has helped their students and made them better educators. Their wisdom and advice is worth heeding. Listen, learn, and apply their wisdom in your classroom, and you will see changes in yourself, your teaching, and your students.

Making a Case for a Research-Based Teaching Rationale

Mrs. Ramsey was an outstanding fourth-grade teacher. Everyone looked forward to being in her class—it was fun, we learned a lot, and we knew she liked us. We thought it was just her personality that made the class so enjoyable. But it was not just her personality—Mrs. Ramsey had a plan and her plan included goals for us and a clear understanding of her role in the classroom.

Built on years of experience, her plan grounded her, providing her with a daily image of what she wanted to achieve in the classroom. Each day she could visualize her success based on her plan. While she had little formal education research to draw on, she had a long history of continual thinking about teaching and systematic observations of students. Knowing her students, understanding her role, and developing a plan enabled her to achieve her vision.

Read and study this book, use its ideas wisely, and it will help you as a science teacher achieve the same kind of success as Mrs. Ramsey. Throughout this book we demonstrate the importance of developing a detailed plan or rationale for teaching science that will help you achieve improved results in the classroom. Although every teacher's plan or rationale will be different in some ways, we have found, based on more than 45 combined years of helping teaching candidates and teachers develop their own teaching rationales, that most successful plans embrace ten key components. The most useful and effective plans

1. are written down and express a personal vision of success in the classroom;
2. fit the personality, goals, and experience of the teacher;
3. are flexible and subject to change;
4. are based on research, not just instinct or personal experience;
5. set specific goals and expectations, and define the desired roles for both students and the teacher;
6. identify and adopt successful strategies to use in the classroom, and build on past successes;
7. seek to create a positive environment for learning in the classroom and aim to meet the individual needs of students;
8. assess results and performance on a regular basis;
9. select content that meets standards while also encouraging creativity; and
10. are developed and shared with other teachers.

Some Research About Exemplary Teachers

For many years educators have been studying exemplary teachers, trying to determine the nature of their excellence and, with this knowledge, to help others become equally excellent at promoting student learning. One seminal study took place 20 years ago and its results still have relevance today. Between 1982 and 1987 the National Science Teachers Association (NSTA) conducted a nationwide Search for Excellence in Science Education. Teams in every state, organized by Robert Yager of the University of Iowa, identified several hundred exemplary school science programs. As the teams studied and wrote about the programs, they also studied the teachers (at the end of this chapter, see sample titles from the Focus on Excellence series, 17 volumes published by NSTA that reported results from that study of programs and teachers).

Without fail, the teachers in this select group, those who designed and implemented wonderfully successful programs, were just like Mrs. Ramsey—each had a plan and great insight into how to make their classrooms embody the plan. Not surprisingly, their students had amazing success at learning (Bonnstetter, Penick, and Yager 1983). While many of the Search for Excellence teachers were "naturals"—born, it seemed, to be excellent teachers—we also know that we can help others be just as successful. This book is designed to do that.

In related studies, Joseph Krajcik, a former president of the National Association for Research in Science Teaching and now at the University of Michigan, studied seven years of teacher education graduates from the University of Iowa (Krajcik 1986; Krajcik and Penick 1989). Each graduate had spent considerable time developing a written, research-based rationale for teaching and had defended it orally several times. Krajcik found that many of these new teachers demonstrated characteristics almost identical to those of expert, experienced teachers as characterized by several large, national studies (Weiss 1978, 1985). As a result, he concluded that we did not have to depend on "born" teachers; rather, teachers could be made—if their teacher education included systematic development of a well-researched plan that guided their teaching and their students' learning.

Reasons for a Rationale

While all teachers have reasons for their actions, some may base their actions more on intuition, hearsay, or even misguided experience. Many have been encouraged to develop a "philosophy of education" made up of the ideas of well-known educators combined with personal feelings about teaching and learning. In this book, we go a step further, differentiating between a "philosophy" and a "rationale." A rationale is built on research, while a philosophy, although built on a thoughtful and intellectual base, may place little emphasis on research support for what the roles of teachers and students should be. Although research in education is not as pervasive (or persuasive) as might be desired, the purposeful teacher should use all available knowledge to inform his or her teaching.

A key to success is to put your plan in writing. Most successful teachers we have observed have written rationales that go well beyond the normal lesson plan. These plans specify their goals and expectations, what content they want to cover, their views of their own teaching roles, and their assessment strategies. The most effective teachers justify and support their ideas with research, thus creating a research-based rationale to guide teaching. Randy Smasal, a veteran high school teacher and now an administrator, recalls how he found little satisfaction in teaching until introduced to the idea of a written rationale (see sidebar, p. 8).

Teachers face a daunting task as they work with a wide array of students, with multiple needs, in physical facilities that are often far from ideal. In this environment, a research-based rationale acts as a blueprint that gives the teacher an overview while maintaining perspective and intellectual control. Having a blueprint and a sense of control reduces anxiety and builds self-confidence (Druckman and Swets 1988).

We also see the development of a personal teaching rationale as the cornerstone of personal professional development during a teacher's career. Just setting goals is not enough. You must construct a plan for growth and change. Developing a research-based rationale can be quite useful for your own success and recognition as an educator. Having reasons, justifications, for your teaching also provides survival value, helping you stay on track with your ideas, plans, and actions.

The Importance of Goals

Effectiveness in teaching means having and achieving specific goals for students. (See more on goal setting in Chapter 3.) Goals are both the beginning and the end of successful teaching. Knowledge and implementation tie all the pieces together. A well-developed personal teaching rationale includes setting goals and defining objectives to reach those goals. Effective teachers do not pick goals out of a hat or have them imposed on them by administrators. They consciously develop their own.

Penick and Bonnstetter (1993), during a 10-year period in more than 25 states and countries, asked parents, teachers, administrators, and scientists about their goals for K–12 students in science education. Regardless of the group, all responded similarly. They wanted their students to

- become more creative,
- be effective communicators,
- use science to identify and solve problems,
- know how to learn science, and
- develop a positive attitude toward science.

Most of these goals involve processes—how you approach learning—or personal characteristics more than specific objective knowledge. Teaching only for content cannot achieve these broad goals. Effective teachers can and do teach mandated content and skills—but they do so in a way that is consistent with their overriding goals, such as those above.

Educational goals are developed by thinking about what knowledge, abilities, and dispositions students should acquire or develop during their K–12 years. Our educational goals for students usually include not only subject matter concepts but also technological and scientific literacy, lifelong learning, inquiry, creative thinking, communication, application of science in a social context, critical thinking, nature of science, problem solving, social growth and development, appreciation of diversity and equity, and development of a worldview, among others. Each goal requires attainment of a variety of skills, knowledge, or attitudes, often in a particular sequence. Thus, our goals are built on a foundation that must be developed one stone at a time (Bransford, Brown, and Cocking 2000).

Of course, the best-laid plans often go astray. Effective teachers know this, plan for it, and are ready to follow an alternate path when needed. Like contingency plans in any critical environment, a rationale provides a vision of the desired outcome and alternative courses of action to reach that goal. With a rationale for their work, teachers are better prepared for a variety of eventualities.

Regular reflection on one's goals is a must. Daily, we think about how and what to teach tomorrow. At year's end, we reflect on the goals attained and set new goals and ideas for the following year. Reflecting is not an idle activity; it, too, must be sequenced and organized. A carefully designed rationale includes elements that provide this organization. When such reflection is done well, we are rewarded with ever increasing success in our classes.

> *The cycle of research and reflection is never ending because with every question I work to answer, several others pop up. I have to maintain my focus. In order to do this, I have to apply what I learned in my undergraduate years through the research-based rationale experience. This experience allows me to focus by identifying goals and then researching and developing actions for both my students and myself. This allows me to continue to seek the "absolute truths" in what I am doing in the classroom.*
>
> —Brandon Schauth, Second-Grade Teacher, Iowa, 2003

Ken Tangelder, a seventh-grade life science teacher in New York, saw an opportunity regarding his goal of using technology in the classroom:

> When I first started teaching, the school had just opened a new computer lab. I can remember our very first faculty meeting when our principal said, "Get those kids down there; that thing is open before school, during school, and after school. We want them down there as much as possible. We want them to be there as much as they are in the library." I said, "Okay," and right off the bat I started looking at one of my three goals, which was to implement technology. How can I use technology in the classroom? Well, the principal had served it to me on a platter. Not a problem; I signed up for as many times as there were available and I went from there. (2003)

Without a reasoned set of goals, would Ken have seen the opportunity so quickly? Would he have moved as fast, becoming the first teacher to use the facility? Or would he have hesitated, perhaps so long that there was no longer room for his classes? One wonders whether, if he didn't have a rationale or that particular goal, he would have gone on to develop the successful activities that are now used by all the teachers at his level. While one cannot do controlled research about such questions, we can state that this type of success is a common story even among novice teachers with whom we have worked when they have solid, well-supported rationales for their teaching.

Value of Research

Research gives us assurance of validity and reliability and guides daily practice as well, helping teachers focus on their goals rather than being sidetracked by interruptions and trivia. Research is the foundation for knowledge and reflection. Without research, actions become guesses and learning becomes a mystery.

> Effective teachers can describe eloquently and in detail how they make links between research and practice, filling in gaps with logic, experience, new knowledge, and even research of their own.
> —Jennifer Rose, Middle School Science Teacher, Minnesota, 2003

With strong research grounding (as will be discussed more fully in Chapter 3), the teacher follows a consistent pattern of seeking effective and proven avenues for goal attainment while minimizing trial and error. These teachers approach teaching as scholars—looking at the classroom as a dynamic and human system of causes and effects—rather than as technicians fixing one small problem at a time.

Research also can help define the benchmarks for a teaching rationale by identifying standards and best practices. National and state standards are common benchmarks frequently incorporated into teachers' rationales. Standards provide specific information for content, pedagogy, programs, systems, and assessment and are used in developing assessments for state programs. While the various state standards and frameworks and the National Science Education Standards (NRC 1996) vary considerably, all tend to focus on students learning through inquiry, on a select set of concepts to be taught, on appropriate sequencing of topics, and on valid assessments. Having a plan helps to carry out the standards. Thus, a research-based rationale aligns standards, benchmarks, and activities with goals, producing a coherent blueprint for classroom action and success.

> *Developing a personal teaching rationale is an important opportunity to contemplate and continue to develop a description of the science teacher you wish to become. Developing this rationale is important because the type of teacher you want to be will strongly influence the methods and techniques of teaching that you will employ in your classroom. A rationale will encourage thinking in very specific terms about the kind of classroom you will have; the ways you will interact with your students, parents, colleagues, and administrators; and the teaching tactics you will emphasize. All these factors will shape the science teacher that you become.*
> —Robert Horvat, Teacher Educator, Ohio, 1999

As teachers develop their rationales and become more versed in the skills needed for implementation, they tend to become tenacious, assertive, purposeful, and influential. They feel capable, see themselves as leaders, and feel they have much to offer their profession. They also may experience negative feedback from colleagues who see these changes as a challenge to their own way of thinking. However, other colleagues will likely look to them as leaders and sources of knowledge and skill.

My two colleagues saw me working on my rationale; they wanted to know what I was doing and they wanted to be in there doing the same thing. I always get them coming to me for ideas and suggestions from what I have done based on some of the goals and things that I have done in my rationale. But I told them if they wanted it (my rationale), they would have to modify it on their own.

—Ken Tangelder, Seventh-Grade Science Teacher, New York, 2003

How a Research-Based Rationale Influenced My Teaching

My first real teaching experience in a science classroom was a 10th-grade biology class and I was the teacher assistant. It left me empty. I began to think of teaching as the delivery of an unending list of facts and tidbits. I thought of students as sponges whose job was to absorb knowledge. I didn't think I wanted to be a teacher anymore.

For my second teaching experience, I was assigned to student teach biology and chemistry with an enthusiastic and brilliant veteran teacher. My time in his classroom was life altering. He modeled everything that he was trying to teach me. He incorporated so much research on effective teaching into the classroom that I began to see the science of teaching in a different light. I had read articles about teachers and had seen research studies done on teaching, but I never really put that knowledge into action in the classroom the way he did. It didn't take long before my attitude toward and respect for science education research completely changed.

While observing my cooperating teacher, discussing the lessons with him, and reflecting on the research, I began to view teaching as a robust, research-based endeavor. My definitions of teaching and learning began to evolve. I started to teach in a way that was research based. After all, as a novice teacher I had a very limited background of personal experiences to draw from when making curriculum and teaching decisions. My cooperating teacher required me to write a research-based rationale for teaching science, explaining what my goals were for students, how the research supported those goals, and which specific teacher behaviors I would use to push students toward my goals. This was by far the hardest paper I had ever written because it was so reflective of what I was doing on a daily basis. This research-based rationale began to define who I was as a teacher and became my framework for making classroom and teaching decisions.

My cooperating teacher had prepared me well, and with a research-based rationale supporting and justifying my decisions, I felt very prepared to move on and teach on my own. Teaching now seemed pretty exciting and I was anxious to try it on my own.

In my first interview, I was offered a middle school life science teaching position. I taught at that school for three and a half years before moving to a high school to teach physical science and biology. I'm now in my 11th year of teaching. Research on effective teaching has become the backbone of my educational approach. I am constantly reflecting on whether a particular assignment, project, activity, or lesson is meeting my desired student goals. In the future, I hope to work with student teachers who are committed to teaching and have an understanding of and a positive attitude toward the research on effective teaching. After all, this is a large part of what makes teaching a profession.

—Randy Smasal, High School Science Teacher and
Assistant Principal, Minnesota, 2003

References

Bonnstetter, R. J., J. E. Penick, and R. E. Yager. 1983. *Teachers in exemplary programs: How do they compare?* Washington, DC: National Science Teachers Association.

Bransford, J., A. Brown, and R. Cocking, eds. 2000. *How people learn: Brain, mind, experience, and school.* Washington, DC: National Academy Press.

Druckman, D., and J. Swets, eds. 1988. *Enhancing human performance: Issues, theories and techniques.* Washington, DC: National Academy of Sciences.

Krajcik, J. S. 1986. An evaluation of the University of Iowa's science teacher education program, 1977–1984. Doctoral diss., University of Iowa.

Krajcik, J. S., and J. E. Penick. 1989. Evaluation of a model science teacher education program. *Journal of Research in Science Teaching* 26 (9): 795–810.

National Research Council (NRC). 1996. *National science education standards.* Washington, DC: National Academy Press.

Penick, J. E., and R. J. Bonnstetter. 1993. Classroom climate and instruction: New goals demand new approaches. *Journal of Science Education and Technology* 2 (2): 389–395.

Weiss, I. R. 1978. *Report of the 1977 national survey of science, mathematics, and social studies education.* Washington, DC: U.S. Government Printing Office.

Weiss, I. R. 1985. *National survey of science and mathematics education.* Research Triangle Park, NC: Research Triangle Institute.

Representative issues (1983–1986) of the Focus on Excellence series, published by the National Science Teachers Association

Penick, J. E., ed. 1983. Focus on excellence: Science as inquiry 1 (1). 131 pp.

Penick, J. E., ed. 1983. Focus on excellence: Elementary science 1 (2). 157 pp.

Penick, J. E., and R. J. Bonnstetter, eds. 1983. Focus on excellence: Biology. 1 (3). 122 pp.

Penick, J. E., and V. N. Lunetta, eds. 1984. Focus on excellence: Physical science 1 (4). 74 pp.

Penick, J. E., and R. K. Meinhard-Pellens, eds. 1984. Focus on excellence: Science/technology/society 1 (5). 103 pp.

Penick, J. E., ed. 1984. Focus on excellence: Physics 2 (1). 88 pp.

Penick, J. E., and J. S. Krajcik, eds. 1985. Focus on excellence: Middle/junior high science 2 (2). 99 pp.

Penick, J. E., and J. S. Krajcik, eds. 1985. Focus on excellence: Chemistry 3 (2). 49 pp.

Penick, J. E., ed. 1986. Focus on excellence: Earth science 3 (3). 41 pp.

Yager, R. E., and J. E. Penick, eds. 1985. Focus on excellence: Non-school settings 2 (3). 109 pp.

Elements of a Research-Based Rationale

Before we discuss how to create your rationale for teaching (see Chapter 3), let's examine several of the components of a successful rationale. In Chapter 1, we mentioned 10 key aspects that most plans embrace. In this chapter, we will look more closely at how to address these, including setting goals for students and defining the roles of students and teacher in the classroom. In addition, we will discuss classroom environment, content selection, and the need for regular progress assessments.

Goals for Students

All teachers (and nonteachers as well, for that matter) express goals that extend well beyond mere content to be learned. Student report cards have for years had spaces for rating student "conduct" or "citizenship" or "deportment." Yet, rare was the teacher who taught explicitly for this "other side of the report card." While many report cards have dropped these categories for evaluation, teachers and others still describe a group of goals other than content for their students.

Yet, even as teachers express a wide assortment of cognitive, behavioral, and attitudinal goals for their students, observers of teaching regularly report that only one goal is visible in most classrooms, that of learning discipline-centered facts, concepts, and skills. John Goodlad (1983), in his massive study of more than 14,000 classrooms, noted this as one of his major conclusions. It's not that teachers don't want to achieve these other goals; rather, most have not seriously and systematically considered their goals other than the content topics prescribed in curriculum guides or frameworks. And, even if they have pondered their goals, they often don't see the implicit relationship between their goals and the roles of students and teacher.

Role of the Students

Achieving any particular goal requires action, often specific to that goal. With our well-considered goals in mind, and with reference to logic, the literature, and experience, we may predict which action or behavior (the student's role) might result. Regardless of the source of support for a goal, the outcome is the same; if students play this role, follow this action, exhibit this behavior, with time they will tend to move toward the desired goal. For instance, if we want students to be able to apply knowledge, then routine and regular opportunities for applying knowledge must be present in their learning environment. Learning to self-evaluate doesn't happen just by being evaluated by others; students must also do their own evaluations and they must do it often, obtaining feedback regularly (Trowbridge, Bybee, and Powell 2004). Penick and Bonnstetter (1993) envisioned the anticipated role of students in the classroom as being derived primarily from the expected goals or outcomes (Figure 1).

Figure 1. A Three-Part Model Relating Roles and Goals

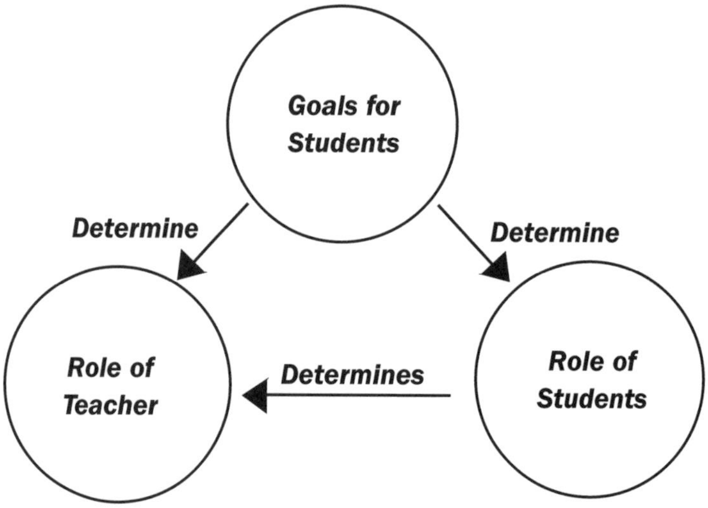

Source: Penick, J. E., and R. J. Bonnstetter. 1993. Classroom Climate and Instruction: New Goals Demand New Approaches. *Journal of Science Education and Technology* 2 (2): 389–395.

Role of the Teacher

The teacher sets the tone of the classroom. Through actions and responses, teachers stimulate what their students do in the classroom. Thus, as you can see from Figure 1, just as the students' roles are derived from the goals we have for them, the teacher's role flows from the desired student roles. In the same way, the teacher's role guides the student role. The teacher's role is to provide materials and learning opportunities conducive to enhancing not just the single goal in mind but the entire goal set. Thus, the teacher's role simultaneously must enhance any particular goal while not endangering other goals. This balancing act also involves using research to plan, describe, and explain which specific teacher behaviors and roles to use in the classroom.

The education research literature contains extensive data specific to some teacher actions or behaviors. Wait-time, that silent and patient pause after we ask a question, is an excellent example of a useful behavior with extensive research backing (Rowe 1974, 1986). If we want students (or anyone, for that matter) to speak or respond to a query, wait-time is not only a necessity, it enhances a broad array of desired student behaviors and, thus, goals. For instance, the data show that increasing a specific instance of wait-time length leads to significantly enhanced and longer student responses, with less hesitation, more confidence and speculation, and more student-initiated ideas. If our goals include or require these student behaviors, then systematically and purposefully we must exhibit wait-time.

> *A teacher's good use of extended wait-time provides many positive effects: increased length of student responses, increased incidence of speculative thinking, increased student-to-student interaction, increased number of student questions, increased use of supporting evidence by students, increased number of students responding, increased student confidence, and increased achievement (Rowe 1974, 1986). Therefore, using longer wait-time is consistent with my goals for students.*
> —Aidin Amirshokoohi, High School Science Teacher, Illinois, 2003

While the need for wait-time is almost intuitive, other common teaching strategies can be counterproductive. Teacher praise of student responses has been researched extensively, and the data consistently suggest that evaluative behaviors (positive or negative) on the part of the teacher can lead to student dependence and conformity by limiting inquiry, thinking, and initiative (Grayson and Martin 1997). Although these consequences are rarely mentioned directly in the goals of teachers, dependence and conformity are the opposites of creativity, which is included in almost all teachers' goals. Virtually all teachers seek to promote student inquiry, thinking, and initiative and many teachers seek for their students to be "independent learners" or thinkers. While few would overtly espouse creating dependencies, teachers must recognize that making evaluations—whether praise or criticism—can inhibit creativity and the learning process.

A research-based rationale guides and encourages you to examine each of your goals and actions logically to see if they have the potential to enhance or obscure your goals and roles. Now, with a clear set of goals and roles, you can begin to develop classroom strategies to meet these goals.

The Classroom Environment

The classroom environment has physical and affective characteristics. Each goal or component of a rationale is best achieved in a particular atmosphere. Teachers with purpose consider what they want to accomplish, looking at how each student's needs can be met by appropriate procedures, by the way the room is arranged, and by the affective qualities that the classroom and they, as teachers, manifest.

In addition, teachers with purpose set up a management plan that facilitates student learning, making information—such as books, media, and web access—readily available. Past assignments and bulletin boards with current work stimulate, celebrate, and promote student thinking. Teachers arrange for group work so that students learn and work together. The intellectual climate and social life of the classroom are modeled by the teacher and complement the physical materials and resources in the room. It doesn't take long to see that you can design your classroom environment a goal at a time, until you have a room designed distinctly for learning.

Selecting Content

Content is rarely mentioned in the initial goals described by teachers. Yet the choice of content to be included in the curriculum is of extreme importance. We must have access to timely and relevant materials at a level conducive to our students' learning. If certain content is incompatible with our chosen goals and classroom environment, however, we must either change our goals or exclude that content. Reconciling these potentially rival aspects is far from trivial and demands a systematic approach. A written, research-based rationale provides for this systematic process.

Over the years, different groups have spent much time and effort trying to determine what science content is most essential. More than 20 years ago, Project Synthesis (Harms and Yager 1981) led to incorporating societal issues and career awareness into science classes. Thinking broadly, the National Science Teachers Association Scope, Sequence, and Coordination project (SS&C) promoted "every science, every year" with a steady building of science knowledge by coordinating the science content to be taught (NSTA 1992). For instance, in teaching about photosynthesis, students would first study the physics of light energy, then the chemistry of light reactions, and finally the biological role of photosynthesis, where the physics and chemistry combine in a biological organism. You might note that this logical presentation of photosynthesis is the reverse of how it is traditionally taught. That was true of many of the concepts studied by NSTA as part of the SS&C project.

Today, the National Science Education Standards (NRC 1996) influence most state curriculum frameworks and publishers, leading to a reduction in the numbers of topics specified in curriculum guides and, to some extent, the numbers of topics in textbooks.

> *Teaching broad unifying concepts that enhance student understanding is my primary focus, rather than isolated facts and trivia. I assess content based on the American Association for the Advancement of Science criteria that take into consideration the personal, societal, philosophical, and intrinsic values the content has for students.*
> —Aidin Amirshokoohi, High School Science Teacher, Illinois, 2003

Assessment

> *If you don't know where you're going, you'll probably get there.*
> —Anonymous

Or, alternatively,

You've got to be very careful if you don't know where you're going because you might not get there.

—Yogi Berra

If getting there is of some importance, then with regularity you take time to ask, "Where am I?" "Where should I be?" "How am I doing?" and "Where to next?" At some point your confidence develops to where you know that if you continue as you have been, you will eventually arrive at your desired destination, your goal.

Our destinations and goals should determine our assessments. If we have a goal, then we should teach for it overtly and, in concert, measure to see where we are in reference to reaching it. If we don't have a goal, we are wandering somewhat randomly with no rational basis for continuing in a particular mode or for continuing at all. When I know where I am going and assess where I am, I never have to wonder, "Am I there yet?" I already know the answer to that one. When our students ask, "Why are we doing this?" or "Am I done yet?" we have clear evidence that they don't know the goals or direction of their (our) classroom journey.

> *Without a research-based rationale I would have followed the example of a few colleagues, which, combined with time constraints of my first few years, would have fixated my attention on multiple-choice tests. The tests would have given me a poor picture of what students knew, where they took wrong turns in their learning, and would not have informed my instruction and planning on a short- or long-term basis.*
>
> *In place of multiple-choice summative tests, I use long answer and problem solving-type questions and grade all of my summative tests with rubrics. This carried over into my instruction, allowing me to convey to students that there are many different ways to solve a problem; indeed, there are many different ways of assessing student knowledge and learning.*
>
> —Brian Fortney, High School Science Teacher, Wisconsin, 2003

Just as your student goals should include more than content knowledge and skills, your assessment domains and tests similarly must include items that deal with all goals, including creativity, problem solving, communication, and so on, in addition to the science content or skill developed. In practice, this should lead to a variety of assessment instruments for measuring attitudes, creativity, skills, problem solving, and any other components found in your rationale. Unless the assessments cover all your goals, then these same goals probably will not be taught for overtly, will not be viewed as important by students and parents, and, in turn, will not be systematically obtained (NRC 1996). If your rationale and goals call for more than science content and skills, your assessments must be authentic to the goals, measuring all of them to the extent possible. If your classroom includes a diverse student body, you may well need a wider range of assessments than is the norm to validly and reliably sample student outcomes. Valid and reliable outcomes require constant consideration of all your goals, not just a few.

Who Assesses in the Classroom?

Traditionally all assessment and evaluation comes from the teacher. Yet, most teachers state in their goals that "students should learn to self-evaluate." Many offer as a rationale the fact that, in the real world, workers are generally expected to know when they are doing something right (or wrong) without waiting for an expert to inform them.

Similarly, all involved in the teaching and learning processes somehow are involved in the assessment of that learning. The teacher's assessment is expected as a professional and as one who has a vested interest in the classroom outcomes. The students' assessment is expected, as students must become more skilled and independent in their thinking, performance, and outcomes. Of course, in order for students to assess validly and reliably they must be taught how to self-evaluate. The teacher with this goal in mind has responsibility for far more than mere assessment of student performance.

Why Do We Assess?

Assessment itself is nonjudgmental, yet judgments are often based on assessments. In the classroom we are faced with a variety of reasons to assess and, ultimately, evaluate and judge. Assessment, while usually focused on students in order to classify them, should go well beyond this single use to include *formative assessments,* which guide instruction as well as assessment of the teacher, the curriculum, the classroom climate, and the overall educational program. In the quote that follows, a teacher asks the questions that are typical of formative assessment.

The research gave me key questions that I ask myself when reflecting on the success of the lesson. I am always asking, What were the students doing today? How can I tell if they were engaged? What did I do to contribute to the success (or failure) of the lesson? How can I teach this differently next time? What kinds of questions was I asking? What kinds of questions were the students asking each other? What kinds of questions were the students asking me?

—Jennifer Rose, Middle School Science Teacher, Minnesota, 2003

Assessing students involves measuring where they are at a moment in time. To do so validly requires having a benchmark against which to measure—a goal or standard that provides uniformity, consistency, and a mark to strive for. Content goals are relatively easy to use as standards as they are somewhat rigid, immovable, and easy to measure. Performance, attitude, or personal characteristic goals, however, require more creativity and finesse to measure. Nonetheless, they should be in our assessment plan as surely as the content assessments are.

Student assessments help us to place students properly in the curriculum or collaborative groups, to determine sequences of instructional events, and to provide for reliable and valid judgments that we may need to communicate to others. Assessments also assist in determining how we, as teachers, are performing. If we truly intend to reach and teach all students, we must know how close we are coming to achieving our goals and where as teachers we have failed or lack knowledge. Without such feedback, we may be committing ourselves to random decisions about teaching and learning or we may be ignoring aspects of education with which we feel uncomfortable. And, obviously, our search for assessment and feedback must be guided by our purposeful rationale.

Assessments also show us how curriculum is functioning and what needs to be changed. We can use assessment techniques to review a prospective curriculum, activity, or material. Without a valid assessment, we are condemned to make decisions based solely on intuition or historical precedent rather than evidence, and we are in a weak position if we must defend our decisions (NRC 2001a, 2001b).

With program assessment we look at the big picture of how our teaching, curriculum materials, and classroom all join to produce a learning environment. Although most teachers do little assessment at this level, this is an area in which we might make the most educational headway if all teachers were guided by evidence and personal, research-based rationales. We can only imagine what would happen if we took the next step, and all teachers in a department, school, or district worked to develop a common research-based rationale and then implemented it broadly and systematically. In this ideal scenario, every hour of every day, all students would encounter teachers with a common purpose and with expectations, skills, knowledge, and strategies to match.

Application in the Classroom

Putting your rationale into action successfully and consistently is the ultimate goal. Executing a plan with precision and élan involves organization and confidence. You need to get off to an appropriate start, sense continuously how students are reacting, and be prepared to change course in midstream or even abandon a particular part of the plan. But, your vision must always be at the forefront.

When things are not going well, the research-based teacher does not abandon her carefully chosen, preset goals. Rather, she modifies style, strategy, or curriculum in some way, usually without breaking pace or disrupting the class.

> *When I first wrote my fantasy scenario, I remember thinking I sounded very naive and idealistic. When I revisited it, I realized that I was idealistic, but not naive, because I didn't sit around and wish for my room to be a certain way, I went out and did it! I begged, schmoozed and borrowed to get the things my students needed, then I set the expectations in my room and dared those kids to meet them. I was amazed to find that by the end of my second year in my current classroom, reality was 80% identical to my fantasy! This was when I realized I had internalized who I wanted to be as a teacher and had subconsciously designed and run my room according to plan.*
> —Deanna Rizzo, High School Chemistry Teacher, New York, 2003

For a more complete story of a teacher who overcame adversity in her classroom by following her rationale, read about Glenda Carter in Chapter 4.

Knowing the elements of a research-based rationale, you now must develop each of these elements and implement them. Read on and learn how. Chapter 3 offers suggestions for developing each element of your rationale, and Chapter 4 provides guidance for integrating your ideas into your curriculum and collaborating with other teachers to help them build rationales as well.

Teaching Rationally: How My Rationale Guides My Teaching

Many students enter my classroom with a generally negative attitude toward science. I tackle this by helping students realize that I have set high yet flexible and attainable expectations for them. I strongly believe that every one of them is capable of learning and understanding science, contrary to what they often believe. The overwhelming majority of my students' attitudes are quickly influenced and change once they realize that with hard work and determination they can meet my high expectations and that I am always willing to provide them the utmost level of support. I could not do this consistently if I did not have a strong, research-based rationale.

My own research-based rationale is based on my student goals that, in turn, are consistent with the current learning theories and effective teaching strategies as indicated by research. Important components in my classroom—such as student actions, teaching behaviors and strategies, classroom climate, content, and assessment—are crucial in achieving these goals. I am a firm believer that the craft of teaching is not so very different from the art of learning. They each depend on curiosity, organized effort, commitment, and the joy of discovery.

In implementing my rationale, I create a comfortable learning environment, which encourages the expression and exchange of ideas, risk taking, critical thinking, creativity, and cooperation. I accomplish this by creating a positive teacher-student interaction by asking stimulating and higher-order critical-thinking questions, allowing wait-time for students to reflect, responding to students in a nonevaluative and nonjudgmental manner, and displaying accepting and empathizing verbal and nonverbal behaviors.

I incorporate several teaching strategies such as the learning cycle, cooperative learning, and science/technology/society—all of which reflect current education research. My students are responsible for and are directly involved in their own learning. Instead of simply being told about a concept, students are involved in their own experiments and activities, helping them see the concepts visually. This visual and hands-on approach has proven to be very successful. The students are often given an inquiry question that they reflect on individually or with their peers and then conduct their experiments or activities to see what takes place and why. The labs are interspersed with various questions that require the students to reflect on the steps of the experiments and end with analysis questions that ask for critical thinking and relating to real-life situations.

For some scientific concepts, students learn through conducting research and presenting their findings to their classmates in very creative ways. They are encouraged to use their creativity by painting; making skits; producing videos; writing stories, poems, and plays; and using various software such as PowerPoint to produce eye-catching, educational presentations. This method helps the students tremendously since presenting the material clearly and creatively to their peers requires them to have a thorough understanding of the topic.

I ask questions that are open-ended and require elaboration. When asking questions, I follow the HRASE strategy (History, Relationships, Application, Speculation, Explanation) to decide on the sequence and types of questions to ask students (Penick, Crow, and Bonnstetter 1996). Questions posed in this sequence encourage students to think in a logical, progressive manner. In addition, using this strategy gives me information about students' thought processes and conceptual understandings, which helps me adapt lessons and activities to students' needs.

After asking questions, I consciously think about using wait-time I (pause after asking a question) and wait-time II (pause after a student's response). Research (Rowe 1974, 1986) and my own experience demonstrate that there are many positive effects of wait-time. These include increased length of student responses, increased incidence of speculative thinking, increased student-to-student interaction, increased number of student questions, increased use of supporting evidence by students, increased number of students responding, increased student confidence, and increased achievement.

The way a teacher responds to students strongly influences the classroom climate. I always try to respond to students in a nonevaluative manner since terminal responses such as criticism or praise generally result in less positive attitudes, increased conformity, increased dependence on others, and lower achievement by students (Costa 1985). Also according to Costa, extending responses that accept rather than evaluate generally result in more positive attitudes, increased risk taking, creativity, self-confidence, and achievement by students. I regularly use other extending responses that include using the student's idea to ask additional questions that require further clarification or elaboration (Abraham and Schlitt 1973). By using extending response patterns, I create an accepting classroom climate where students feel comfortable expressing their ideas to me and other students.

Within this framework, I emphasize learning subject-matter disciplines in the context of inquiry, technology, science in personal and social perspectives, and history and nature of science. The students combine processes of science and scientific knowledge as they use scientific reasoning and critical thinking to develop their understanding of science.

As teachers, we must understand that reflection is an integral part of professional growth and improvement of instruction. My rationale guides me in assessing my own effectiveness as a professional. My data sources include classroom observation, information about students, pedagogical knowledge, and research. I regularly take advantage of video recordings of my teaching to directly observe the classroom environment and analyze student involvement, interactions, and learning as well as my verbal and nonverbal behaviors and strategies. I collaborate and share a variety of instructional resources with my colleagues and collaborate with other professionals as resources for problem solving, generating new ideas, sharing experiences, and seeking and giving feedback. Reading professional publications such as *The Science Teacher, Educational Leadership, NEA Today*, and *Scientific American* and attending professional conferences such as the annual NSTA national convention have contributed greatly to my professional development.

Having developed and used a research-based rationale has led me to yet another goal: I want to serve as a leader for science education reform. Leadership is a requirement for any change. Effective leaders reflect critically on their own practices, continually strive to improve themselves, and are open to new ideas. I will serve as a leader in science education by not being simply a consumer of research, but rather one who will be critiquing prior research, conducting and publishing my own research, and giving presentations at conventions. I want to assist in producing effective future teachers who are ready to adopt new methods or adapt old theories to new classroom realities.

—*Aidin Amirshokoohi, High School Science Teacher, Illinois, 2003*

References

Abraham, M., and D. Schlitt. 1973. Verbal interaction: A means for self-evaluation. *School Science and Mathematics*: 678–686.

American Association for the Advancement of Science (AAAS). 1993. *Benchmarks for science literacy.* New York: Oxford University Press.

Costa, A. L. 1985. *Developing minds: A resource book for teaching thinking.* Washington, DC: Association for Supervision and Curriculum Development.

Getzels, J. W., and P. W. Jackson. 1963. *Creativity and intelligence.* New York: John Wiley.

Goodlad, J. 1983. *A place called school.* New York: McGraw-Hill.

Grayson, D., and M. Martin. 1997. *Generating expectations for student achievement.* Canyon Lake, CA: GrayMill.

Harms, N. C., and R. E. Yager, eds. 1981. *What research says to the science teacher.* Vol. 3. Washington, DC: National Science Teachers Association.

National Research Council (NRC). 1996. *National science education standards.* Washington, DC: National Academy Press.

National Research Council (NRC). 2001a. *Choosing content using national science education standards.* Washington, DC: National Academy Press.

National Research Council (NRC). 2001b. *Classroom assessment and the national science education standards.* Washington, DC: National Academy Press.

National Science Teachers Association (NSTA). 1992. *Scope, sequence and coordination of secondary school science: The content core.* Washington, DC: NSTA.

Penick, J. E., and R. J. Bonnstetter. 1993. Classroom climate and instruction: New goals demand new approaches. *Journal of Science Education and Technology* 2 (2): 389–395.

Penick, J. E., L. W. Crow, and R. J. Bonnstetter. 1996. Questions are the answer. *The Science Teacher* 63(1): 27–29.

Rowe, M. B. 1974. Wait time and rewards as instructional variables, their influence on language, logic, and fate control: Part 1. Wait time. *Journal of Research in Science Teaching* 11: 81–94.

Rowe, M. B. 1986. Wait time: Slowing down may be a way of speeding up! *Journal of Teacher Education* 37(1): 43–50.

Trowbridge, L., R. Bybee, and J. Powell. 2004. *Teaching secondary school science: Strategies for developing scientific literacy.* Upper Saddle River, NJ: Pearson.

Developing a Research-Based Rationale

Developing a research-based rationale requires formulating a way of thinking about teaching. Rather than a casual, weekend effort, the creation of a complete rationale is a long-term enterprise that may last for years and may never be finished (and probably should not be).

In this chapter we offer you the guidance of a number of teachers to help as you select and refine goals, identify roles for students and teachers, describe your desired classroom environment, select content to be taught, and formulate assessment models for your teaching practices and student learning. Finally, we suggest how to develop a base of research findings that support your rationale.

My own research-based framework is based on my student goals that, in turn, are consistent with current learning theories and effective teaching strategies as indicated by research. Important components in my classroom, such as student actions, teaching behaviors and strategies, classroom climate, content, and assessment, which are crucial in achieving these goals, interact as a supporting framework in leading toward such goals.

—Aidin Amirshokoohi, High School Science Teacher, Illinois, 2003

Identifying Goals for Students

The initial process of developing a rationale is not complicated, but it is thought provoking and time intensive. Once developed, however, updating and maintaining a rationale become almost routine.

You can develop a research-based rationale alone or in a group (such as a school department or a professional education class or one made up of interested colleagues). In any of these situations, having a guide or colleagues who work together will enhance building a rationale, as you have much to discuss, consider, and decide. An experienced guide makes the process flow, and discussions with colleagues open up avenues of creativity and inspiration you will appreciate. Rationale development in preservice teacher education courses has been shown to work well, in part because of the multiple opportunities for interaction, others seeking the same ends, and the thoughtful efforts of a skilled instructor. But if you are working alone, don't despair—we've given guidance to teachers in your situation many times before. Read these chapters, think about how this applies to you, and write your ideas down. In the end, you will have a rationale to be proud of.

We agree with experienced teachers and teacher educators who find it best to start by selecting broad goals for students—goals that focus on a broad spectrum of student learning. If you already have a set of goals, revisit them while reading this section, looking at them closely in ways you may not have considered. We'll begin with the three basic steps in the process of setting goals for students:

- Brainstorming
- Refining the goals
- Developing support for the goals

Brainstorming

Goals are our starting point. Until we articulate specific goals for students and commit to them, we can't really design our instructional plan or our curriculum. In our science methods courses, we always begin with an activity to identify these goals. We call this *goal generation,* and we use brainstorming as the central process.

Goal Generation

If you're going to generate ideas, you need an initial starting or focal point. One might begin with Penick and Bonnstetter's (1993) question, "After 13 years of formal schooling in science, what goals would you like your students to achieve?" or "After completing your science program what do you want your students to be like, to be able to do, and to know?" These questions get at the attitudes, dispositions, products, and content knowledge that you think should be part of a purposeful science program.

Brainstorming is a good way to get started. It is a powerful group activity, where one idea leads to another. (For a twist on traditional brainstorming, see sidebar on p. 44, "Negative Brainstorming: A Technique for Finding the Positive.") In this process, we try to let our minds wander, making free associations between thoughts, with little concern for directed thinking. The process, while perhaps more powerful and quicker in a group, can easily be adapted for an individual working alone. Alone, you can write your ideas as a list or an outline, draw pictures, or do whatever else keeps the process moving. This is not a time to reflect, question, modify, or even elaborate on your initial ideas. That will come later (Harris-Freedman 1999).

Two rules keep the brainstorming process strong. First, remember that all ideas are valid and must be added to your list. Write them all down, just as you heard them in the group or thought of them on your own. Second, don't evaluate at this time. Just write down ideas as they come to mind. If someone disagrees with an idea, write down both points of view. Evaluation of ideas limits free association, an essential component of creativity.

The Facilitator's Role

When brainstorming occurs in a group, the facilitator must wait patiently and quietly after the initial focus question. Typically, participants are slow to begin to offer ideas, but as they see their ideas written down without evaluation, the pace will increase (this holds true even when you are brainstorming by yourself). All suggestions should be listed one after the other for everyone to see. A group of preservice teachers typically can generate 30 or so goals in about 20 minutes. There is usually duplication, ambiguity, and overlap. At this point, don't worry about it. We will deal with that in the next phase on page 29. Even if the pace of ideas slows, don't give up. Thinking creatively takes time and effort, and ideas don't come with a uniform pace or style.

If you are the facilitator, you can stimulate or even suggest ideas, but wait a while to make sure you are not overly structuring the activity. If you jump in too fast, many in the group will just let you do the work, and they will watch silently and passively as you create your list, not theirs. If you have been patient, using lots of wait-time, and after 15 minutes or so you still note areas that have been neglected, pose a goal yourself. For instance, you might say, "How about creativity?" and then wait. If no one picks up on it, you now can either take the liberty of adding it yourself (you are a member of the group, after all!), or letting it die. (If you're working alone, of course, you don't have to worry about any of this except writing down your ideas and not rushing to finalize your thoughts.)

Part of your strategy as a facilitator is to be prepared and anticipate what might occur. We routinely find that groups that are generating goals initially leave out not only creativity (which we think is the essence of science), but other concepts like independent thinking (necessary for continued learning and advancement), and, surprisingly, science content itself. If science content is not listed after about 15 items are named as potential goals, ask, "What about science?" This usually quiets the room, as participants scan the list, looking for the science. They often come back with statements like, "Science is in the problem identification and solving," or "Science is in 'Students will apply their knowledge.'"

With this lack of specific content in their specified goals, we like to tease participants a bit by asking, "What about 'Know the periodic table'?" or 'Newton's third law'?" They always tell us these aren't real goals; these are just short-term objectives and not the lasting knowledge or characteristics they want citizens to possess. At this point, some are already questioning what they have traditionally taught or what they plan to teach, as well as how. These questions prompt the group to look at their list one more time, seeking to eliminate trivial goals and to make all listed goals clear and worthy of inclusion.

An individual who is brainstorming alone can follow the same rules as for groups, but extend the writing time, possibly over several days. Whether in a group or alone, you can let your writing sit for an hour or a day or two, then revisit it and add more. Like good soup, sometimes it is better the next day, as it ages and, perhaps, new ingredients are added. Let the first ideas you write down lead you to other ideas. If you do better making a concept map rather than a list, put your thoughts down that way. Or you might turn to your methods text or the national standards documents for inspiration.

Refining the Goals

After all the brainstorming, you probably have a list of disorganized and possibly even contradictory goal statements. Now, it's time to refine each goal. While there are a number of ways to do this, we suggest two mechanisms. One of us (John) likes to follow basic goal generation by breaking the larger group into small groups and providing this challenge:

> *Reduce this list on the board down to a maximum of 15 goals you will take as your own. You may rewrite, clarify, combine, or add to the list.*

About 30 or 45 minutes later, after vigorous debate, each group has its list. This activity allows all to talk in the safety of small groups, propose and defend ideas, and take leadership roles. Many who don't speak out in large groups can be quite vocal in a small group.

Once the small groups each have a list of 15 goals, the instructor brings them back together as a whole and then says,

> *Now, as a whole group, reduce the list still further to no more than 10 goals. You have 15 minutes.*

With that, John leaves the room, forcing leadership to arise within the class. As individuals have been debating in small groups, they have arguments ready for the large group. Usually there is enough similarity in the group ideas that only four or five of the goals provoke most of the debate.

When the students finish, they announce with obvious satisfaction that they have their 10 goals. By this time, they are usually committed to these goals, as they have been debating and defending them for almost an hour. They have faith in these goals and they have talked about them enough to understand what is meant by each. They do not let go of these goals easily and they usually want to describe and explain them to others. This is as planned.

Robin likes to follow the brainstorming by breaking the large group into small groups and saying,

> *Take a look at the list of ideas and begin to sort them into groups by writing each idea on a "stickie" or card for sorting. Remember that ideas may fit into more than one group.*

The use of white boards or butcher paper with markers is a quick way to make the sorting visible for all. Each group then shares its sorting and titles with the large group. They present an oral version of how they picked the groups and titles they used in their sorting. Students reveal their prior knowledge, misconceptions, and differences in understandings. Most importantly, all get to express their ideas, disagreements, and suggestions in a safe environment.

If you are by yourself, you can write each idea on a card or stickie and spread them out on a table where you can see them all. Think or talk aloud about how each of the ideas might relate to one another. If you made a concept map to begin with, revisit the connections. Start to move the goals into several groups, maybe four or five, and then give each group of goals a general title. This title should identify the common characteristics of the goals you placed here. Sometimes a brainstormed goal becomes the title for a group. Remember that if a goal fits in more that one group, just make an extra stickie or card and place it in another group.

Developing Support for Goals

Once goals have been established, a thorough and persuasive rationale includes a justification of each goal. This justification typically demonstrates the value of this goal for the individual and society, sometimes through logic, philosophy, or belief and sometimes from research. For instance, "Students will exhibit creativity" might be justified philosophically by noting that variety and aesthetics are pleasing, part of every culture, and viewed by most as desirable. From a research perspective, creativity might be defended by noting that creative people have been shown to be more observant and more dependable, to have better health, and to learn and retain more information (Barron 1963). From a purely logical point of view, we might justify creativity as necessary for continual progress and improvement of products, processes, and life itself. Creativity might even be a goal for all educators as highly creative students score higher on standardized examinations than do students with low creativity scores (Getzels and Jackson 1963).

Finally, many will justify a particular goal by looking to the discipline itself. For instance, in teaching science, all the standards-related documents (such as the *National Science Education Standards* [NRC 1996]) seek for our students to understand the nature and history as well as the concepts of science. For creativity, we might appeal to statements from scientists and philosophers such as, "Creativity is the essence of science," or "Innovation and discovery are two sides of the same coin," or even "Science is going beyond the information given."

Obviously, the goal that can be justified though all of these avenues—such as "Students will exhibit creativity"—is among the easiest to defend and support and probably is open to multiple mechanisms for incorporating into the daily curriculum. Yet, being easy to justify does not mean it automatically will be easy to apply in the classroom, to develop in students, or to assess. And just because it is hard to implement is no reason to discard it as a goal. In many ways it's a lot easier to be a teacher without purpose, although not very effective, efficient, or satisfying to teacher or students.

Developing a Research Base

With details of pertinent aspects of the goal, we now move to the literature of research related to the goal. But before you dive into the educational research databases, you might organize the essential elements of each goal by writing a paragraph or two about each, derived from your personal experience and knowledge (Rutherford and Ahlgren 1990; Weld 2004). Remember, though, some of your personal thoughts might include misconceptions about what is good teaching. While developing a research base to support each goal, you may have to let go of some of your prior ideas about effective teaching.

Begin your search for evidence with the ideas you have identified logically as potentially viable and valuable. We suggest starting with the resources included in this book or a database such as ERIC (*http://www.eric.ed.gov/*). In the process, you are going to move from just being able to say, "This is how I teach" to being a purposeful teacher who says, "and this is why, and what I expect to happen." As you find credible sources and read them, they will lead you to yet other sources. Look in journals that report educational research. All of the National Science Teachers Association (NSTA) member journals *(Science and Children, Science Scope, The Science Teacher,* and *Journal of College Science Teaching)* contain research articles in each issue. NSTA members can access all four journals online, as well as topic indexes, at *www.nsta.org/publications.* (Also be sure to see the Appendix on p. 47, "Rating Credibility of Research Sources," for a method that can be used to evaluate research. The scoring guide in the Appendix is especially useful for web-based material.)

Research journals such as the *Review of Educational Research* often have concise review articles that summarize a wide selection of prior research publications and are rich in information. These always include numerous references to other articles you might find useful. Some of your goals may even be represented by a complete journal, such as the *Journal of Creative Behavior.* In addition, recent books, such as *Classroom Instruction That Works: Research-Based Strategies for Increasing Student Achievement* (Marzano, Pickering, and Pollack 2001), offer research-based strategies on topics such as cooperative learning and generating and testing hypotheses. Two other sources of research-based information for educators are *Tips for the Science Teacher: Research-Based Strategies to Help Students Learn* (Hartman and Glasgow 2002) and *Educating Teachers of Science, Mathematics, and Technology* (NRC 2001). Spend regular time in the library adding to your knowledge and understanding of the education research and you will be well rewarded.

Research about education provides power and enthusiasm for teaching, as these teachers make clear:

> *My passion for teaching stems from the research I have had the opportunity to grapple with. I have found that the more secure my teaching philosophy is, the more accurate my intellectual autonomy is to judge the research. Asking questions of the research you read leads to yet another source of research.*
> —Brandon Schauth, Second-Grade Teacher, Iowa, 2003

> *For the past 17 years, I have indexed the journals I receive (through*

the efforts of student assistants who type the article titles into a database for me). I keep the journals in my storeroom, and can do keyword searches on topics and retrieve lists of articles on subjects related to science teaching. I regularly search the database and read or file the articles and "how to" ideas in my course guide books as references and activities. I think a teacher could teach an entire high school course out of the wealth of information published in journals over the years.
—Paul Tweed, High School Biology Teacher, Wisconsin, 2003

Identifying Desired Roles for Student and Teacher

Now that you have goals that are well described, carefully considered, and justified by research, you are ready for the next step: determining the necessary roles of students and teacher to achieve them. Identifying roles is the part of rationale development where you translate what research says into purposeful practices. You will now answer questions such as, "What will my classroom look like and be like?" "What are my expectations for my students?" and "What will I be doing when I teach?"

We now turn our focus to *how* to structure our teaching, curriculum, and classroom climate to achieve our goals. There are multiple ways to approach this task. And, as Figure 1 (p. 12) indicates, our student goals provide a path and connections between student and teacher roles. Connecting all the paths shows clearly the relationship of goals and roles.

Although there are many ways to isolate and describe the student and teacher roles that will most likely lead to the desired goals, we suggest that with each goal you write out the key concepts and your personal understanding of the goal. For example, pick your first goal—"Students will use scientific methods of inquiry"—and write out what you know are the essential elements of inquiry. Using the National Science Education Standards (NRC 1996), for example, you see that inquiry has fundamental understandings, abilities, and processes. From other literature, such as Reif, Harwood, and Phillipson (2002), Robinson (2004), and Harwood (2004), you will find that questions are a key component of inquiry. From your intuition and logical understanding of appropriate roles, continue on to seek research support for each goal.

We find that a graphic organizer, such as the Role Identification Matrix (Table 1), which was developed by Robin's class, can be useful in arranging and coordinating your ideas related to goals and roles. The focus questions within the matrix should start you thinking about how you will achieve a particular goal, how students will learn, how you will teach, what the classroom climate will be like, and how you will know you are successful. As an example, the matrix in Table 1 identifies the goal "Students will be creative" (Marks-Tarlow 1996; von Oech 1998) and shows how one group elaborated on this particular goal.

Role of the Student

Begin by writing in the first column of the matrix what you would expect to see students doing if they were being creative. With the creativity goal in mind, our first thought might be that different students are doing different things. There would be unusual activity, things that you as teacher might not have predicted. Students should be expressing their creativity, perhaps talking with each other or by producing a creative product. Rather than just following teacher directions, some students will be making their own decisions, structuring their own activity, lesson, or learning.

Even if they are all doing the same experiment or activity, some will approach it one way and others in a different fashion. Students' comments will vary. Some students will be thinking as you do and others will not. There will be much speculation about classroom events.

As you work with your goals, you will gradually expand on them, eventually developing a catalog of descriptions detailing your actions and those of your students. Second-grade teacher Craig Leager expanded the student roles column for his written rationale (Figure 2). This list, slightly condensed from his original, provides an idea of how much detail you can create just by thinking about goals and roles.

Role of the Teacher

Now, with the student column filled in, return to the teacher column of Table 1. With each envisioned role of the student, there should be a corresponding role for the teacher. Sometimes you will think of something you *do* and sometimes something you should *avoid* doing. But you must consider carefully *how* you will go about teaching to obtain the desired student outcomes you listed in the first column (Yager 1991; Shapiro 1994; Rakow 1996/2000; Texley and Wild 1996/2004; Lowery, Texley, and Wild 2000; Haley-Oliphant 1994; Trowbridge, Bybee, and Powell 2004).

Table 1. Role Identification Matrix

Goal: Students will be creative.			
Role of the Students *Focus question:* What will my students do and be like?	**Role of the Teacher** *Focus question:* What will I do and be like? List specific behaviors and actions.	**Classroom Environment** *Focus questions:* How will I create the class environment? What will I do to develop the affective environment?	**Assessment** *Focus questions:* How will I know I am successful? What will students' products be?
Will not all be doing the same thing.	Am creative in my lesson design by having a number of different activities available.	Display creative student work, encourage and celebrate student differences.	Not all products are identical.
Will have opportunities to structure class events and make decisions.	Stay current by reading and researching about creativity. Incorporate new ideas into classroom instruction.	Display pictures, posters in the classroom, changing with new events and issues. Make it look like a science room.	Students begin to bring in relevant and interesting materials and ideas.
Will express unusual ideas. Not all students will have the same ideas.	Am accepting of unusual and different ideas from students.	Create a safe work space, intellectually and physically.	Discussions are varied, lively, and unexpected, with unusual points of view.
Will approach experiments, activities, and ideas from different directions.	Use creative ways to resolve real situations, showing examples and posing problems.	Provide a wide variety of resources to encourage student use of variety of information.	Students find more, possibly better, solutions, as well as new problems.
Will answer frequently with depth and variety.	Ask open-ended questions and use appropriate wait-time.	Offer more independent investigation; all investigations are conducive to inquiry.	Students are learning through inquiry, not just *about* inquiry.
Will be open in their answers, offer speculation, and try new things.	Do not evaluate student responses. Limit evaluation in general and teach students to self-evaluate.	Make the overall climate intellectually comfortable, safe, and nonthreatening.	Students disagree with me and others, have their own opinions and support them.

Figure 2. Craig Leager's Student Goals and Actions

Goal: Students are critical thinkers.
Student Actions:
Articulate and attempt multiple solution paths
Ask and answer questions
Reflect and evaluate their thinking processes
Make connections across tasks
Debate and challenge ideas
Assess the credibility of sources
Relate their own personal experiences to learning situations
Solve problems
Goal: Students have a "voice" in their learning environment.
Student Actions:
Verbalize their opinions, questions, and concerns
Assist in the development of classroom rules
Aid in issue resolutions with peers/groups
Suggest topics of interest for study
Share personal experiences
Goal: Students take intellectual risks.
Student Actions:
Question concepts and content
Seek alternatives to given approaches
Communicate suggestions (verbally, in writing)
Debate and challenge ideas
Goal: Students are resourceful learners.
Student Actions:
Ask questions of teachers and peers about concepts and content
Use print and technological resources to answer questions
Search for information using resources beyond those at home or school
Use information from personal experiences to make connections
Investigate suggestions made by others
Goal: Students use problem-solving skills.
Student Actions:
Attempt problems using various approaches
Seek advice and guidance from others to solve problems
Test and modify ideas
Look for relationships between ideas and topics

Figure 2. *(continued)*

Complete activities and tasks
Goal: Students make cross-curricular connections.
Student Actions:
Draw on personal experience for information
Use resource materials from one subject area in another
Communicate similarities and differences of concepts and content between different subject areas
Ask questions about relationships among subject areas
Goal: Students demonstrate cooperative behaviors and exhibit social competence.
Student Actions:
Offer assistance to peers/others
Mediate resolutions with peers/others
Maintain a clean and orderly work area
Listen to one another's suggestions, comments, and questions
Offer and accept constructive criticism
Goal: Students are self-motivated learners.
Student Actions:
Communicate opinions and concerns
Attempt problems without directly being instructed
Use problem-solving techniques
Set personal goals related to one's own education
Goal: Students communicate their own ideas effectively.
Student Actions:
Participate in class
Make eye contact during conversation
Work cooperatively
Share personal thoughts related to topics of discussion
Ask questions
Goal: Students are creative.
Student Actions:
Hold many ideas at once
Have and share original ideas
Listen to and analyze new views
Attempt multiple solution paths
Be independent in one's thinking
Take risks

For instance, if you want students each to do something different, you might command, "Everyone must do something different!" Unfortunately, creativity does not occur on demand. And a command from the teacher is in many ways the antithesis of what one expects in an open and creative environment. So, in this case we might write in the teacher role column, "Avoid unnecessary directions," as directions can be quite constraining. But, of course, we must eventually figure out what we *will* do. In this case, rather than "Avoid unnecessary directions," we might phrase it as, "Keep directions as minimal as possible," or "Structure directions so that students must make many decisions as they complete the assignment."

So, what are the actions of a teacher in a creative classroom? Some research has found that modeling creative behavior is an asset in teaching others. "Model creative behavior" is still rather vague, however. But if we think about what would we be *doing* if we were modeling creative behavior, we can come up with more specific actions, such as

- Using multiple ways of expressing ideas or solving problems.
- Accepting and valuing diversity of thought and action.
- Continuing to learn, becoming able to respond to student ideas more easily and with yet more ideas.
- Asking more open-ended questions as opposed to those that require a simple or memorized response.
- Asking questions to gain new information rather than to test student knowledge or skills.

The list could (and should) go on. Do this with each of your student goals, and soon you will have a set of specific behaviors and roles that make up the heart of your teaching rationale.

The Classroom Environment

If you can picture something in your mind, understand it, and refine and revisit the picture often, you can begin to make it real. In the case of your science classroom, "making it real" means creating an environment in which the students respond to your role as you envisioned it in the second column of Table 1. For the third column, think about the classroom setting where you and your students spend the day. Imagine this setting with no monetary, resource, or space barriers. Think of what you would like it to look like physically. Now, bring the teacher and students into the picture, pick a content topic and imagine the live action in your classroom. What physical characteristics of your room would help you achieve those goals you have established? How are materials arranged and accessed? What stimulating items are present? How are students and teacher positioned in the room? (Motz, West, and Biehle 1999).

In addition to physical aspects, your room has an emotional climate. You create that climate through your behaviors and role. You can easily imagine your role if you wished to scare all your students, to make them feel incapable of learning, unworthy of respect, or even bored. Just as your role and the role you allow and expect for students can lead to these negative feelings if you create a negative climate, so can your role lead to a positive environment (Darling 1993; Garcia 1998; Harris 2004). When they were preservice teachers, all of the teachers whose comments appear in this book wrote imaginary scenarios that described their ideal classroom environments and the actions of the students and the teacher that would create those environments. These scenarios, while imaginary, were firmly rooted in careful planning, research, and analysis. Moving from the imaginary to the real requires no more than thoughtful and skillful implementation in your classroom.

Furthermore, you create more of the intellectual climate of the classroom through implementation of your role than with the physical materials and resources in the room. What you expect from students, the freedoms you allow, how you respond to students, and your overall demeanor often are far more influential than the mere objects in the room. You have a lot of power and it can be used to further your goals or to inhibit them.

Assessment

The fourth column of the Role Identification Matrix is "assessment" (Table 1). Asking certain focus questions—for example, "How will I know I am successful?" and "What should my students be doing or learning as a result of my actions?"—will help you identify the specific types of assessment you can use to evaluate you and your students' actions (Doran et al. 2002; Darling-Hammond 1993).

While assessment is the last column and usually the last activity of a teaching unit, it must be considered prior to and during instruction, not just after. Assessment informs you, guiding your teaching while alerting you to both successes and problems. Assessments must be developed as you develop curriculum and must be a key feature of your research-based rationale. In a well-designed rationale, the goals, the roles, and the assessment are all consistent and compatible. If you don't plan for this, you might easily have assessments that counter your prior activities or even your goals. End your teaching with assessment, but don't leave it for last.

Assessment of Students

Assessments are used to classify students, but they must go beyond this single use to include formative assessments that guide instruction (Atkin and Coffey 2003). If we truly intend to reach and teach all students, we must know how close we are coming, which goals are showing progress, and where we as teachers have failed or lack knowledge. Without such feedback, we are left to random decision making or we may be ignoring aspects of education with which we feel uncomfortable.

Self-Assessment by Teachers

In addition to assessment of students, teachers with purpose use their rationales to look inward, routinely asking themselves questions such as

- How do my classroom practices align with my rationale?
- What components of my rationale are not being addressed in my practice? When and how is this occurring? What can I do to change this?
- How do the science learning opportunities in the classroom match my rationale?

- In the context of my rationale, what appears to work and what does not appear to work? What needs to be revised and how would I do it?
- What new understandings have evolved from my teaching experiences?
- What revisions are needed for my rationale and why?
- What do I need to learn?

Once you have described your role, how could you not consider how your actual role matches your desired role? Perhaps the most valuable aspect of your rationale is a clear image of what you want to be doing as you teach so you can compare this image to your actual teaching performance assessment.

> *I constantly reflect on my position in the class. Were higher-order questions used appropriately? Were questions phrased so as to maximize student synthesis of knowledge? Did my questions elicit student conceptions of what was being taught? This is a constant battle, and seems, at times, daunting. Over time, reflection on my questioning techniques served to improve my perception of what students were learning, as well as my use and phrasing of higher-order questions.*
> —Brian Fortney, High School Science Teacher, Wisconsin, 2003

Many teachers videotape themselves in the classroom and then use some system of objective analysis for coding and assessing their behaviors and roles. Evaluation comes into play when you compare the observed (the assessment) with the expected performance and ask basic questions such as, "How are these different?" "How important are the deviations and how do I explain them?" "How must I change to obtain the performance I desire?"

Airasian and Gullickson (1997) provide eight reasons why self-evaluation is an important process for teachers. They note that self-evaluation

- Is a professional trait and responsibility.
- Focuses professional development and improvement on the classroom or school level where teachers have their greatest expertise and effect.
- Recognizes that organizational change is usually the result of individuals changing themselves and their personal practices, not

of "top-down" mandates.

- Gives teachers voice—that is, a stake in and control over their practices.
- Builds awareness of the strengths and weaknesses of personal practice; it grows from the immediacy and complexity of the classroom, as do teachers' motives and incentives.
- Encourages ongoing teacher development and discourages unchanging classroom beliefs, routine, and methods.
- Improves teacher morale and motivation.
- Encourages collegial interactions and discussions about teaching.

Obviously, all professionals self-evaluate and teachers should be no different. If you must rely on others for your professional evaluation, you are forever dependent on their points of view and you must wait to make changes until the external evaluator is present. Imagine driving your car and having to wait for a backseat driver to tell you when to slow down or to turn. A truly professional teacher makes expert observations, assessments, evaluations, and decisions based on a solidly grounded rationale for teaching. Table 2 is an example of a reflection matrix for asking questions in the classroom.

Bringing It All Together

Now comes the fun part, as you implement your well-developed research-based rationale for teaching. Chapter 4 guides you in implementing and developing your own skills while offering still more suggestions for developing, understanding, and using your rationale.

Purposeful teachers' evidence and insights come from regular and systematic study of themselves and their classrooms, making them action researchers in settings where instructional problem solving takes on new meaning. Research involves asking questions, seeing problems and opportunities, trying new ideas, and experimenting with new ways of teaching and structuring the classroom. These teachers use their personal rationales as frameworks not only to teach, but to study what is working or not working as desired.

Table 2. Reflection Matrix on the Use of Questioning

Research-Based Statement From Rationale	Evidence From My Classroom Practice That Supports This Statement	Reflective/Evaluative Comments About My Classroom Practice	What Action Do I Plan to Refine My Practice?
Student learning and performance is better when I ask good questions—that is, questions that are consistent with my desired goals.	Students seem to match my behavior. Students with questions show more curiosity and interest in the lessons. Interested students stay on task better. I have prepared questions prior to teaching lessons. I ask questions of varying cognitive demand. I have videotaped myself teaching and used a checklist to evaluate my question asking. I practice wait-time. I ask students to monitor my question-asking and to provide feedback.	The focus on the kinds of questions that I ask has helped me in several ways. I have been able to ask questions that have led to a higher quality of classroom discussion. Students have become more comfortable with responding and are more engaged in what we are doing. I have been surprised by how deeply some of my students can think. Students appear to provide better responses on test questions that call for synthesis of information.	I plan to continue what I am doing, since I am a work-in-progress in this area, but I have begun to internalize this process. I plan to continue to have students involved in this process. As students move through the year and become familiar with levels of questions, I could have them write questions requiring more cognitive demands. I also plan to use a closure assessment in which students write down three things they learned, two questions they still have, and one thing they understood for the first time. I'll see how this goes and make modifications as needed.

Source: Designed by Sandra Enger, Teacher Educator, Alabama, 2003

As a teacher/scholar, I am revisiting the research and adding/deleting parts of my rationale. My plans should grow with me, right? I always want to learn new things—I buy books, read articles, and talk to other teachers. As a teacher/researcher, I want to find out how I can be better. I also want to know what is that something that makes one teacher able to reach students, another one fail. Is it just having written a rationale, or is it internalizing it and implementing it that makes the difference?

—Deanna Rizzo, High School Chemistry Teacher, New York, 2003

Negative Brainstorming:
A Technique for Finding the Positive

As a high school senior, I took a mandatory class called "Communism Versus Democracy." While much of the course focused on events of the first half of the 20th century, there was also a decided flavor of indoctrination. Many students felt this and reacted negatively, even though we all much preferred what we knew of democracy to our visions of communism.

One day our teacher was focusing on the evils of war. For the first time I recall in that class, he had us do a brainstorming activity. Actually, it was one of the few times we had done an activity. As such, it was a welcome change of pace. The teacher proceeded to the board and asked us to list the positive and negative aspects of war. At first it went slowly and then, for some reason, the pace picked up. We began to have fun, much to the teacher's chagrin, as we found more and more positive reasons for war. Soon our list looked like this:

ASPECTS OF WAR

Positive	Negative
We get more land	They lose land
Employment increases	People die
Men look sexy in uniforms	It costs a lot of money
New technology is developed	People are mad at each other
The men left at home have more dates	
We get to use our army	
Our army gets real experience	
War makes other issues seem less urgent	
War leads to great songs and movies	
War makes heroes	
We get to demonstrate bravery	

Of course, our teacher did not like or want this list; he wanted one with a lot of negative aspects and few positive ones. We were not very obliging.

Years later, I was teaching an inservice class on developing creativity. I had asked the group of teachers to brainstorm all the ways they could think of to enhance creativity. I waited and wrote and waited still more. Yet the list was painfully short. Then, in a moment of inspiration, I recalled my high school class and said, "OK. What if you wanted to kill all vestiges of creativity? What would you do in your classroom?" With that, the teachers seemingly rubbed their hands with glee and generated a long list ranging from "lecture all the time" to "you will be evaluated positively only if your solution looks like mine" to "punish all who are different." We quickly had a respectably long list. Then I said, "If this is what you would do to discourage creativity, what would you do to encourage it?" Without hesitation and almost in unison, the class said, "Do just the opposite!"

With that, I discovered the power of what I call *negative brainstorming*, a technique I have used often. It seems people sometimes find it easier (or more fun perhaps) to think of the negative consequences prior to the positive.

—*John Penick, Teacher Educator, North Carolina, 2005*

References

Airasian, P. W., and A. R. Gullickson. 1997. *Teacher self-evaluation tool kit.* Thousand Oaks, CA: Corwin.

Atkin, J. M., and J. E. Coffey. 2003. *Everyday assessment in the science classroom.* Arlington, VA: NSTA Press.

Barron, F. 1963. The need for order and disorder as motives in creative activity. In *Science creativity: Its recognition and development,* eds. C.W. Taylor and F. Barron. New York: John Wiley and Sons.

Darling, C. 1993. A new attitude: Teachers confront biases in sex equity training. *Vocational Education Journal* 68 (3): 18–21.

Darling-Hammond, L. 1993. Setting standards for students: The case for authentic assessment. *NASSP Bulletin* 77(556): 18–26.

Doran, R., F. Chan, P. Tamir, and C. Lenhardt. 2002. *Science educator's guide to laboratory assessment.* Arlington, VA: National Science Teachers Association.

Garcia, R. L. 1998. *Teaching for diversity.* Bloomington, IN: Phi Delta Kappa Educational Foundation.

Getzels, J. W., and P. W. Jackson. 1963. *Creativity and intlligence.* New York: John Wiley and Sons.

Haley-Oliphant, A. E., ed. 1994. *Exploring the place of exemplary science teaching.* Washington, DC: American Association for the Advancement of Science.

Harris, R. L. 2004. Developing student centered classrooms. In *The game of science education,* ed. J. Weld. Boston: Pearson.

Harris-Freedman, R. L. 1999. *Science and writing connections.* White Plains, NY: Dale Seymour.

Hartman, H. J., and N. A. Glasgow. 2002. *Tips for the science teacher: Research-based strategies to help students learn.* Thousand Oaks, CA: Corwin Press.

Harwood, W. 2004. An activity model for scientific inquiry. *The Science Teacher* 71(1): 44–46.

Lowery, L., J. Texley, and A. Wild. 2000. *NSTA pathways to the science standards: Elementary edition.* Arlington, VA: National Science Teachers Association.

Marks-Tarlow, T. 1996. *Creativity inside out: Learning through multiple intelligences.* Menlo Park, CA: Addison-Wesley.

Marzano, R. J., D. J. Pickering, and J. E. Pollack. 2001. *Classroom instruction that works: Research-based strategies for increasing student achievement.* Alexandria, VA: Association for Supervision and Curriculum Development.

Motz, L., S. West, and J. Biehle. 1999. Science facilities by design: Learning and teaching in science. *The Science Teacher* 66(6): 28–32.

National Research Council (NRC). 1996. *National science education standards.* Washington, DC: National Academy Press.

National Research Council (NRC). 2001. *Educating teachers of science, mathematics, and technology.* Washington, DC: National Academy Press.

Penick, J. E., and R. J. Bonnstetter. 1993. Classroom climate and instruction: New goals demand new approaches. *Journal of Science Education and Technology* 2: 389–395.

Rakow, S. J., ed. 1996/2000. *NSTA pathways to the science standards: Middle school edition.* 2nd ed. Arlington, VA: National Science Teachers Association.

Reiff, R., W. S. Harwood, and T. Phillipson. 2002. Scientists' conceptions of scientific inquiry: Voices from the front. Paper delivered at the Annual Meeting of the National Association for Research in Science Teaching, New Orleans.

Robinson, W. R. 2004. The inquiry wheel: An alternative to the scientific method. *Journal of Chemical Education* 81(6): 791–792.

Rutherford, F. J., and A. Ahlgren. 1990. *Science for all Americans.* New York: Oxford University Press.

Shapiro, B. 1994. *What children bring to light: A constructivist perspective on children's learning in science.* New York: Teachers College Press.

Texley, J., and A. Wild, eds. 1996/2004. *NSTA pathways to the science standards: High school edition.* 2nd ed. Washington, DC: National Science Teachers Association.

Trowbridge, L., R. Bybee, and J. Powell. 2004. *Teaching secondary school science: Strategies for developing scientific literacy:* Upper Saddle River, NJ: Pearson.

von Oech, R. 1998. *A whack on the side of the head.* New York: Warner Books.

Weld, J., ed. 2004. *The game of science education.* Boston: Pearson.

Yager, R. E. 1991. The constructivist learning model: Toward real reform in science education. *The Science Teacher* 58(6): 52–57.

Appendix: "Rating Credibility of Research Sources"

How do you determine the credibility of your sources? In Robin's graduate seminar, students developed a rubric to check the credibility of sources. Table A.1 defines the four criteria they used to determine credibility: Reflects and Supports My Topic, Internal Credibility, Data Support Claims, and Web Credibility. As students read research papers, they looked in different sections for internal credibility. For instance, in the Background section they looked at what claims were being made. In the Methods section, they looked at the description of the sample population and how it was chosen. They also looked at what and how data were collected and the limits of each study.

Next they looked for reliability and validity information. In the Results sections of articles, they looked at levels of statistical significance and whether the data made educational sense. Finally, in the Discussion section they looked to see if the inferences made were connected to the claims and were reasonable given the sample population, the data collected, and the statistics used for analysis. In using this scoring guide, students in this research course were able to sort their references into useable categories. Publications that scored high they used for major supporting ideas. Those that scored lower, if used, were in a minor supporting role.

Writing justifications becomes easier when you have gathered several papers that support your ideas about a goal. For each goal, bring together your original writings, including written statements about the key ideas of each goal. Provide support by linking each key idea to evidence found in the research. Have courage about dropping misconceptions or unsupported ideas you may have about what is good teaching. Now you should be letting research, not just intuition or even experience, guide your thinking. This is the essence of a research-based teaching rationale. Teachers with purpose are those who can and do support their ideas with research.

Table A.1. Credibility Scoring Guide

Criteria for Determining Credibility	Credibility Level			
	High 1	2	3	Low 4
Reflects and Supports My Topic	In the title and throughout the article.	In sections.	Mentioned.	Not related.
Internal Credibility	Written by established expert in the field. Rigorous data analysis.	Peer reviewed. Cites established experts.	Journal or organization has a positive reputation in the field.	No evidence presented.
Data Support Claims	Statistically significant results. Generalizations to other populations adequate and explained.	Methods, what kind of data are collected, reliability and validity information.	Mismatch between claims and data.	No claims or no data.
Web Credibility	Online peer-reviewed journal or publication.	Reputable sponsor. References national organizations such as ACS, AAPT, NSTA, NARST, ASTE, AERA, NAGT, NABT, NESTA[a].	Associated with college or university.	Found in/on Amazon.com, local newspaper, internet, or web.

[a] American Chemical Society, American Association of Physics Teachers, National Science Teachers Association, National Association for Research in Science Teaching, Association for Science Teacher Education, American Educational Research Association, National Association of Geoscience Teachers, National Association of Biology Teachers, National Earth Science Teachers Association.

Source: Modified from matrix developed by graduate students at Buffalo State College, spring 2003.

Implementing Your Rationale and Becoming a Mentor

I n this chapter, we discuss how you can implement your rationale in your classroom and use it to become a mentor to others. As you develop and implement a research-based rationale for teaching, you will find yourself changing. You will teach and communicate with others more effectively and achieve better results in the classroom. This chapter describes how you can implement and make these changes systematically and inspire others to do likewise.

Implementing Your Rationale

We focus here on four core aspects: design, implementation, assessment, and evaluation.

Science Program Design

In designing a science program to implement your rationale in the classroom, you must consider national standards, district and state frameworks and requirements, standards for best practices and assessment, and the resources available. In addition, each individual must look at her or his own skills, as well as the skills of her or his students. Once again, teachers must consider how they will help each and every student to advance to all the stated goals. Such determination entails examining student backgrounds and characteristics as well as understanding all the variables, including the availability of time and resources (Rakow 1996/2000; Texley and Wild 1996/2004; Lowery, Texley, and Wild 2000; Garcia 1998; Hart 1983).

When you incorporate these components into a well-developed rationale, you have a guide for planning the best possible science program. While achieving all of your goals is the ultimate aim of teaching, any given lesson will generally stress just a few goals. But the individual goals provide convenient scaffolding for curriculum and student development since individual goals can be examined and thought of separately. Thus, we design our plan—our rationale—one goal at a time. With each goal, we can determine the role of the students and the teacher, what materials we will need, and what content to emphasize.

After the second week of his third year of teaching, a seventh-grade teacher put it this way:

> First I thought about trying to tackle one of those goals. I thought that maybe taking all at once might have been a bit much for me. So I started off looking at one goal, "Students will use and be comfortable with technology," because that was something being pushed at the time, and I said, "This is a great place for me to start." I began incorporating a lot of technology, not only for me but for my students—getting them involved in more online resources using technology as a tool rather than as a crutch, as it is for a lot of people. And I use it myself as a way to take notes and have students take notes. I use it for kids as a learning experience where they can transmit other information. How do I know it works? Well, looking at what I've done, probably the best way I know it works is to see the students actively engaged in what they're doing and enjoying it.
> —Ken Tangelder, Life Science Teacher, New York, 2003

With the goals, research, and strategies from your rationale, you can look at each aspect of your classroom—such as texts, materials, instruction, environment—and make informed decisions. For instance, selection of a textbook can be based on an evaluation of how it relates to your goals and desired student activities rather than assuming that the latest text is better than what you have.

Your rationale guides design of instruction as you again consider what you would like to see students doing as they learn. Visualizing your students succeeding in the classroom can be a critical part of achieving that success. How can you best develop that vision? In Chapter 3 we introduced a Role

Identification Matrix (p. 35). Creation of your own matrix is an essential part of your design planning; without a matrix it will be difficult for you to be truly purposeful or systematic in your approach. Your analysis of your completed matrix provides you yet more opportunities for thinking about and understanding your vision of teaching and learning.

Using your rationale and completed matrix, you can screen and review topics or activities, asking, "How will this particular topic or activity allow my students to work toward the maximal number of my goals?" or "How is this consistent with my goals and strategies?" or "How might this impede our progress?" You have made a plan and now you work the plan.

Program Implementation

Implementation of the research-based rationale takes time, reflection, skill, practice, and revision. Rather than living with anxiety, a purposeful teacher approaches a classroom with confidence, knowing her or his thinking is grounded in research and reality and will lead to success for all.

> *My students are made fully aware of the goals that I have created for them, as well as the actions that should accompany each goal. I believe that this empowers students and gives them reason to strive for success. Parents have shared their pleasure with me in relation to having specific goals and actions that reach beyond a mandated curriculum and speak to the humanity of each student's personal experiences.*
> —Craig Leager, Second-Grade Teacher, Iowa, 2003

And when things don't work out as expected, the teacher-with-rationale looks inward, examining why. Rather than discard the goal (or blame the students) this purposeful teacher returns to her roots, her carefully planned rationale. She might say, "If the goals are right and the research is right, then it can be done."

Glenda Carter returned to teach middle school after a number of years as a teacher educator at North Carolina State University. In the sidebar "A Teacher Educator's View From the Classroom Window" (p. 55), she describes how her rationale kept her going, even when problems arose.

Program Assessment

Just like the feedback you collect from students, data on your teaching can be gathered over several iterations of the curriculum. When you have three or more opportunities to teach the same material each day, your observations of yourself and your students (especially when you compare them to your desired images) easily guide adjustments that maximize your educational impact. No longer random or unsystematic, this type of reflection and revision ensures a positive classroom evolution.

Coaches operate in much the same way as they watch videos and analyze the game, noting where improvements can be made (Weld 2004). So it is in the assessment and evaluation of a program plan. Your rationale, in conjunction with informal and formal student assessments and teacher reflections, must guide changes in your program. Once you have taken action in the design, implementation, and assessment of your teaching program, you will find yourself changing for the better.

Each week compare where you are with where you want to be. You are on the threshold of becoming a purposeful teacher when, rather than reacting to what happens in your room, you consider, reflect, and make modifications that keep the class moving, in unison, toward the goals you have set.

One way of checking to see if you are moving in a positive direction toward your goals is to analyze your teaching through peer review or videotape analysis.

> *I regularly take advantage of video recordings of my teaching to observe the classroom environment and qualitatively and quantitatively analyze student involvement, interactions, and learning as well as my verbal and nonverbal behaviors and strategies.*
> —Aidin Amirshokoohi, High School Science Teacher, Illinois, 2003

> *I make a point to watch other teachers teach. I systematically study the behaviors of the teachers and students. I gather answers to my questions about teaching and learning.*
> —Jennifer Rose, Middle School Science Teacher, Minnesota, 2003

Evaluation

While assessment is no more than a measure, evaluation places a value judgment as to the worth or value of that particular measure. Teachers with purpose continuously assess, and then evaluate, usually by comparing actual measures to those desired (Airasian 1994). A research-based rationale provides a mechanism for objectively and systematically evaluating instruction, student learning, the curriculum and classroom climate, and even your rationale.

> As a classroom teacher I use active research as a way of refining my own teaching. Simple yet effective tools such as pre- and posttests and surveys work very well in developing a feedback system for the effectiveness of a class or program. For many years I have used these techniques not only for content-oriented information, but for attitudes toward science and the environment, measures of creativity and the ability to connect and apply science knowledge and methods to out-of-class situations, and student problem-solving abilities.
> —Paul Tweed, High School Biology Teacher, Wisconsin, 2003

Becoming a Collaborator and Mentor

Once you have an image of who you are as a purposeful teacher and have a rationale to guide your practices, you can begin to share your ideas and thinking with others. More so than teachers without a rationale, almost all teachers with rationales say that they are comfortable when asked to demonstrate or justify their teaching practices. In fact, many look forward to evaluation visits from school administrators or to showing their skills by presenting at professional meetings, teaching demonstrations, and workshops.

> I invite administrators and peers into my classroom so that I can gain their insight and perspective. I make it a habit to invite my principal to visit during those most productive days. I am not afraid to keep my classroom doors open. Teachers need to be able to share their practices with others.
> —Jennifer Rose, Middle School Science Teacher, Minnesota, 2003

Effective communication requires at least two people; to be effective, each party must use and understand the same language. A research-based rationale facilitates accurate communication by providing a common tongue—that is, common ideas and language that focus on research and observable events in the classroom. As teacher educators, we feel strongly that a teacher is not fully expert unless he or she is able to describe and explain his or her rationale to other teachers.

> *I noticed that most people (including administrators) don't seem to be able to articulate what an effective science classroom looks like. I use my understanding of research to explain myself to anyone who might question what I am doing. What's right for science classrooms according to evidence from research may not always be what is familiar to most people. In fact, most parents and administrators seem completely unaware of what constitutes effective science teaching. When I discuss what I'm doing as a teacher and explain the reasoning behind it with the parents of my students, it gives them confidence that their child is in good hands.*
> —Jenni Geib, High School Physical Science Teacher, Missouri, 2003

Not surprisingly, professional teachers with rationales, expertise, and high levels of comfort in explaining their roles and research are often found to be collaborative and to be mentors of others. They can describe and explain—making their visions, actions, and skills clearly understandable to others. They are teacher educators as well as teachers.

> *As a mentor, I try to befriend those who are newer than me or just seem to be struggling. I don't know if it's the rationale, the process we went through when developing it, or both that pushes me to share what I know/learned with others. I always ask myself, Does this person have a rationale? Has he or she ever thought about writing out a formal plan for teaching? This is something I encourage someone to do.*
> —Deanna Rizzo, High School Chemistry Teacher, New York, 2003

An effective and professional teacher has strong visions and missions related to classrooms and teaching. Such visionary teachers become leaders, viewing themselves differently because of their knowledge of what good teaching is and because they have the skills, knowledge, and dispositions to realize their visions. They can conceptualize and explain their visions to others. Like Glenda Carter (see sidebar below), they do not give up just because it doesn't work the first time. They use their rationales to guide what teaching becomes and to make their dreams of an effective classroom real.

> *Preservice teachers at the local university always ask me, What are the most important things in science teaching? I tell them to get connected to the science teaching community at the state and national level, read journals, and use the ideas that fit your situation. Have a well-thought-out rationale for what you do as a teacher. Collect all the resources you can that help you understand your area and effective teaching strategies. Be flexible and take risks to improve the opportunities for your students.*
>
> *Always stand up for your program, promote student work, write articles, invite the media in, get students involved in the community, beg, borrow, and write grants to get equipment so you have the best science lab in your area. Show students you are interested in your field, that you are interested in them as students and as members of your community. Be curious; never stop learning. Use your own ignorance as a tool to show students that learning never stops. And most of all, have a sense of humor and enjoy what you are doing.*
>
> —Paul Tweed, High School Biology Teacher, Wisconsin, 2003

A Teacher Educator's View
From the Classroom Window

Like many of my science education colleagues, I had several years of precollege teaching experience, albeit almost 20 years ago. Although I knew the research literature and was cognitively convinced of the efficacy of research, I wanted to apply some of the strategies in context and for an extended period of time. So, in the spring of 1998 I accepted an interim position at a middle school where student teachers were often placed. I taught full-time from March 9 to the last day of school, June 5, providing me an 11-week experience as a full-time teacher.

I had four heterogeneous classes of sixth-grade general science and two periods of electives—animal science and remedial reading. I decided to focus on three aspects of reform-based middle school teaching: an inquiry approach to learning science, the use of cooperative learning groups, and the implementation of alternative assessment strategies including journal writing, performance assessment tasks, and open-ended questions. Since I was familiar with most commonly cited barriers for not using these constructivist based strategies, I thought I would explore constraints of time and the science curriculum, classroom management, established culture of the science classroom, and personal identity issues.

Initially, I thought that time barriers and covering the content would not be an issue. Philosophically, I wholeheartedly subscribed to the "less is more" emphasis of the National Science Education Standards. I felt no personal or professional pressure to finish the text or defined curriculum and was skeptical of the commonly offered platitude that all the content had to be covered because the students would need the information for the next year. However, by April, I was questioning instructional decisions that lengthened the instructional process. By the end of the year I even questioned that trade-off of breadth for depth. My journal entries reveal much of my thinking at the time.

April 14

I am having the greatest difficulty deciding how much freedom to give students to make mistakes. It's quite a dilemma for me to make an instantaneous decision on whether or not I am going to allow students to go down a pathway that will not get them where I would like them to go. This is much more demanding and I am conflicted much more often than I ever was when teaching traditionally. I am tempted almost every day just to give them the answer. It is a constant struggle to stay in this process. And I am only struggling with myself.

Struggling with the conflicts between my research-based rationale and my efforts at implementation was exhausting.

May 6

I am really tired. I persist in this very draining approach only because I am firmly convinced that its best practice as defined by research. I can't imagine how I would persist if I hadn't engaged in studies that gave me the opportunity to look at teaching in depth. A research paper can never capture all the complexities of the reality and richness of the teaching and learning endeavor. Understanding the complexity is probably what motivates me to continue tweaking, to align the theory and practice.

I had freedom to do what I wanted and I didn't have to worry about evaluators coming in or about getting tenure or being rehired. I didn't have to worry about the other teachers in the department. Although I had anticipated that parents and students would not provide constraints to teaching science constructively, what I hadn't anticipated was why. I quickly realized after attending the first few parent-teacher team conferences that I was unlikely to face any constraints or complaints because science and social studies were not tested. Even with my own view, I kept seeing this experience through the eyes of the new teaching graduates who may enter a teaching position in the middle of the year.

March 11

The science storeroom is a dump for mostly outdated and broken equipment. And only the science department chair is permitted to have a key to the storeroom.... I thought it "amusing" that the science department head walked into my room after school today and pointed out that I had two microscopes and two balances on my back table that were not supposed to be there. First she asked where they came from and after I told her that they had been there when I arrived, she said, "They belong in the storage room."

I gained a lot of new respect for what my preservice students face when they try to implement research-based teaching strategies.

April 20

Today I went against what I know is best practice and what my own research has indicated. I actually put three low-level students together to work on the lab. Even as I was doing it I was arguing with myself, "No. Wait. Figure this out." The other part of me said, "I can't, I have to do this now, I will think about this later." I am bothered by what I feel is defaulting to what is easier for me.

And it didn't happen just once.

April 23

OK, I have done it again. I pulled Benny totally away from the group and I am making him work alone. Of course I am giving Benny points for completing his work. This is not what I wanted to do. How can I go against my own research findings? I know that research always has to be reworked in context. But I am frustrated because I don't consider this to be an unusual context. If I were less experienced though, I might feel that it was just another case of theory and practice just not meshing. I don't really feel that way.

After the first rush of excitement at having worked through multiple roadblocks, a set of fears, which appear in retrospect to be very rational, assailed me. *Suppose I fall flat on my face. Suppose I can't do what I sometimes so blithely advocate? If I were to fail at this task, what would I do for a living?* I was not surprised that the results of implementing some of the recommendations did not go as smoothly or as well as I would have liked. However, I managed to persist in the implementation even in the face of failure. As I reflected on my own willingness to do this, I came to realize that this persistence arose as a direct result of my knowledge of the research literature and my firmly entrenched belief that research informs best practice. I knew the research was sound and educationally it made sense. Somehow, I just was not able to put in into practice yet. My rationale, so carefully developed over a number of years, provided me a faith to carry on, to find a way to make it work.

This trust in educational research makes me markedly different from many preservice and inservice teachers. Not only do they usually lack necessary skills, they may give up on implementing innovative research-based strategies because of distrust in educational research itself. My own trust in the process was a direct result of my experiences as an educational researcher.

I came to the realization that literal interpretations of research literature may usurp teachers of professional judgment. I speculated that the gap is really one of understanding how the findings of research should be used. That is, using research findings within the context of a particular class means the ideas are manipulated and rotated until a fit with the particular class is possible. This requires considerable skill (and probably experience).

As I recognized that I was in a daily conflict with myself over recommended practice and implementation of that practice, it became apparent to me that I now perceived a phantom gap between theory or research and practice. I realized that part of this gap is a product of producing teachers who are not researchers. Without the research portion of my own understanding, the outcome of my experience would have been much different. I also realized that students need to be engaged in recognizing research findings as guidelines and need practice in framing these guidelines within a classroom setting; part of the profession of teaching is the daily resolution of the conflicts between perceived theory and practice. Learning to teach means, among other things, learning to apply your own knowledge as you resolve the issues that arise routinely. Having a research base for your own understanding is a necessary part of this knowledge.

—Glenda Carter, Teacher Educator, North Carolina, 2003

References

Airasian, P. 1994. *Classroom assessment.* New York: McGraw-Hill.

Garcia, R. L. 1998. *Teaching for diversity.* Bloomington, IN: Phi Delta Kappa Educational Foundation.

Hart, L. A. 1983. *Human brain and human learning.* Oak Creek, AZ: Books for Educators.

Lowery, L., J. Texley, and A. Wild. 2000. *NSTA pathways to the science standards: Elementary edition.* Arlington, VA: National Science Teachers Association.

Rakow, S. J., ed. 1996/2000. *NSTA pathways to the science standards: Middle school edition.* 2nd ed. Arlington, VA: National Science Teachers Association.

Texley, J., and A. Wild, eds. 1996/2004. *NSTA pathways to the science standards: High school edition.* 2nd ed. Washington, DC: National Science Teachers Association.

Weld, J., ed. 2004. *The game of science education.* Boston: Pearson.

Index

*Page numbers in **boldface** type indicate figures or tables.*

intellectual climate of, 39

technology in, 6

Classroom Instruction That Works, 32

Collaborative teachers, 53–55

Communication, 54–55

Costa, A. L., 22

Creativity, 30–31, 34–38, **35–37**

Credibility of research sources, 47

Credibility scoring guide, **48**

D

Designing a science program, 49–51

E

Educating Teachers of Science, Mathematics, and Technology, 32

Educational Leadership, 22

Emotional climate of classroom, 39

Environment of classroom, 14, 22, 25, **35,** 39

ERIC database, 31

Evaluation, 53

Exemplary teachers, 2–3

F

Formative assessment, 17–19

Fortney, Brian, 16

G

Goals, 4–6, 11–12

assessing achievement of, 16–17, **35,** 40

brainstorming of, 27–29

content, 18

developing research base for, 31–33

developing support for, 30–31, 47

generation of, 27

justification of, 30–31

refining of, 29–30

reflection on, 5